2002 99/21

EXPLORING
JUPITER

By Richard Alexander

KidHaven
PUBLISHING

Published in 2018 by
KidHaven Publishing, an Imprint of Greenhaven Publishing, LLC
353 3rd Avenue
Suite 255
New York, NY 10010

Designer: Deanna Paternostro
Editor: Vanessa Oswald

Photo credits: Cover, back cover, pp. 5, 11 Vadim Sadovski/Shutterstock.com; p. 7 PlanetUser/ Wikimedia Commons; p. 9 NASA images/Shutterstock.com; p. 13 Nostalgia for Infinity/ Shutterstock.com; p. 15 Dorling Kindersley/Getty Images; p. 16–17 Victor Josan/Shutterstock.com; pp. 19 (main), 21 Stocktrek Images/Getty Images; p. 19 (inset) NASA/SCIENCE PHOTO LIBRARY/ Getty Images.

Library of Congress Cataloging-in-Publication Data

Names: Alexander, Richard, author.
Title: Exploring Jupiter / Richard Alexander.
Description: New York : KidHaven Publishing, [2018] | Series: Journey through our solar system | Includes bibliographical references and index.
Identifiers: LCCN 2017003818 (print) | LCCN 2017006214 (ebook) | ISBN 9781534522916 (pbk. book) | ISBN 9781534522510 (6 pack) | ISBN 9781534522794 (library bound book) | ISBN 9781534522657 (eBook)
Subjects: LCSH: Jupiter (Planet)–Juvenile literature.
Classification: LCC QB661 .A435 2018 (print) | LCC QB661 (ebook) | DDC 523.45–dc23
LC record available at https://lccn.loc.gov/2017003818

Printed in the United States of America

CPSIA compliance information: Batch #BS17KL: For further information contact Greenhaven Publishing LLC, New York, New York at 1-844-317-7404.

Please visit our website, www.greenhavenpublishing.com. For a free color catalog of all our high-quality books, call toll free 1-844-317-7404 or fax 1-844-317-7405.

CONTENTS

GIGANTIC JUPITER

Jupiter is the largest planet in the **solar system**. It's also the fifth planet from the sun. The planet is so big it can fit 1,300 Earths inside of it!

Jupiter is two and a half times bigger than all the other planets in the solar system put together.

MOVE IT!

Like all the other planets, Jupiter **orbits** around the sun. However, it takes the planet 12 Earth years to orbit around the sun once. The planet orbits slow but spins very fast. It spins all the way around in less than 10 hours!

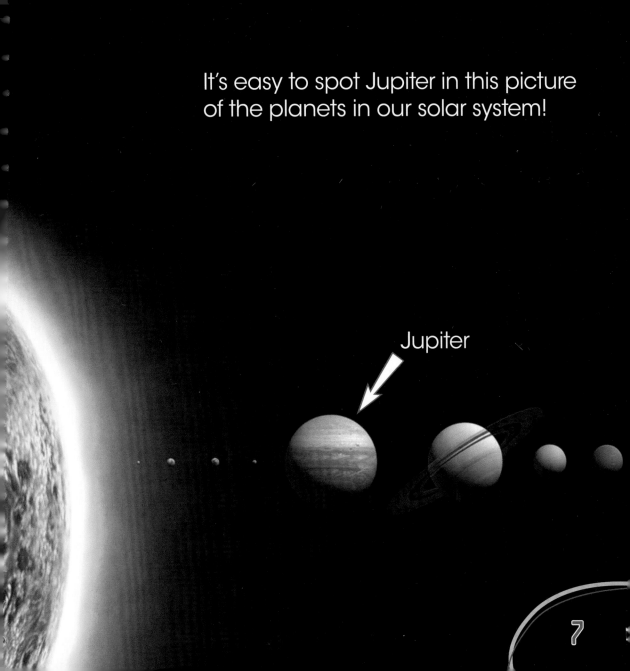

It's easy to spot Jupiter in this picture of the planets in our solar system!

Jupiter

WINDY WEATHER

Jupiter has very windy weather and is covered with many clouds. These clouds are blown around by the wind in different directions. This makes Jupiter look like it has stripes.

Jupiter's **clouds** are made up of crystals of frozen **ammonia,** which create what's called "diamond rain."

There are many storms on Jupiter. The area called the Great Red Spot is a storm that's lasted for hundreds of years. It's the biggest storm in the solar system! Three Earths could fit inside the storm at its widest point.

The color of the
Great Red Spot
changes from
brick red to brown
because of the
ammonia crystals
in Jupiter's clouds.

Great Red Spot

PLANET OF GASES

Most of Jupiter, especially its outer **layer**, is made up of gases. The planet's two main gases are **hydrogen** and **helium**. Unlike Earth, it doesn't have solid ground to walk on.

Jupiter is one of the planets known as gas giants.

Liquid hydrogen and liquid helium are two of the **elements** found just under Jupiter's gas clouds. Jupiter's **core** layer is hot and most likely made of rock.

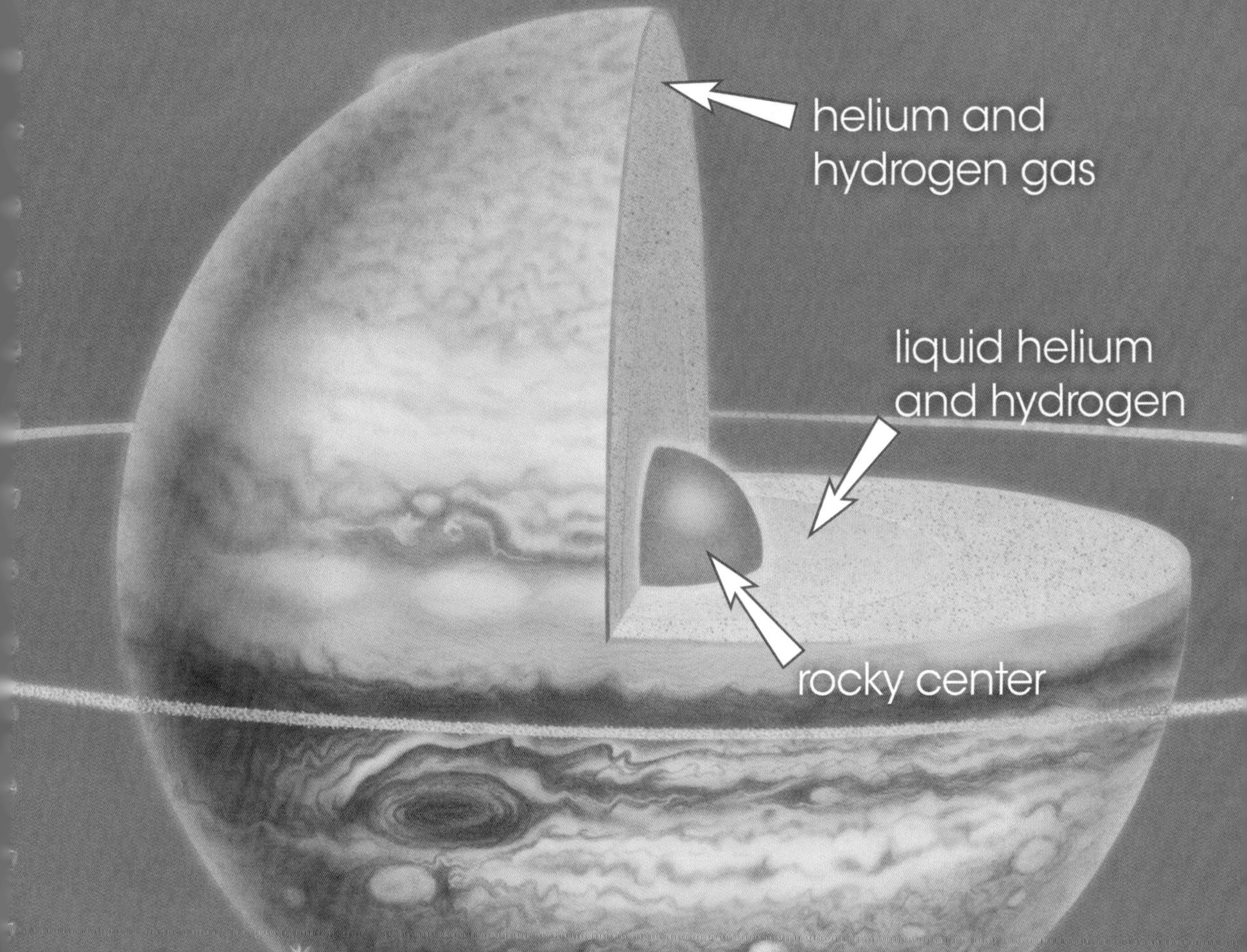

helium and
hydrogen gas

liquid helium
and hydrogen

rocky center

People need to study Jupiter
more closely to be sure what its
layers are made of. Shown here
is what many believe based on
what they know right now.

JUPITER'S MOONS

Jupiter has more than 60 moons. The planet's four largest moons are Io, Europa, Ganymede (GA-nih-meed), and Callisto.

Callisto

Astronomer Galileo Galilei discovered Jupiter and its four largest moons in 1610.

Ganymede

Ganymede, which is bigger
than the planet Mercury,
is the largest moon in the
solar system.

Europa

Io

Jupiter's Europa is a frozen moon. Scientists think it has a liquid ocean underneath its icy **crust**. They also think the ocean could be home to living things.

Jupiter →

Europa →

Jupiter's Io moon, shown here, is home to many **volcanoes**.

Io

19

JUPITER'S RINGS

In 1979, a space **probe** took photos of rings around Jupiter. These rings are made of tiny pieces of dust. They are much harder to see than Saturn's famous rings.

Jupiter's rings can
be seen here.

GLOSSARY

ammonia: A colorless gas with a strong smell.

astronomer: A scientist who studies different parts of the solar system.

core: The center of a planet.

crust: The outer shell of a planet.

elements: Chemical properties that make up the universe.

helium: The second lightest gas in the solar system.

hydrogen: The lightest gas in the solar system.

layer: One part of something lying over or under another.

orbit: To travel in a circle or oval around something.

probe: A vehicle that sends information about an object in space back to Earth.

solar system: The sun and all the space objects that orbit it, including the planets and their moons.

volcano: An opening in a planet's surface through which hot, liquid rock sometimes flows.

FOR MORE INFORMATION

Websites

NASA: Jupiter
www.nasa.gov/jupiter
NASA provides news about and pictures of Jupiter.

National Geographic Kids: Mission to Jupiter
kids.nationalgeographic.com/explore/space/
mission-to-jupiter/#jupiter-planet.jpg
This website features useful facts about Jupiter.

Books

Adamson, Thomas K. *The Secrets of Jupiter*. North Mankato, MN: Capstone Press, 2016.

Owen, Ruth. *Jupiter*. New York, NY: Windmill Books, 2014.

Roumanis, Alexis. *Jupiter*. New York, NY: AV2 by Weigl, 2016.

INDEX